The INSIDE GUIDE

THE MICROSCOPIC WORLD

Minerals

Under the Microscope

By Jennifer Lombardo

Cavendish Square

Published in 2024 by Cavendish Square Publishing, LLC
2544 Clinton Street Buffalo, NY 14224

Website: cavendishsq.com

Cataloging-in-Publication Data

Names: Lombardo, Jennifer.
Title: Minerals under the microscope / Jennifer Lombardo.
Description: Buffalo, New York : Cavendish Square Publishing, 2024. | Series: Inside guide: the microscopic world | Includes glossary and index.
Identifiers: ISBN 9781502667984 (pbk.) | ISBN 9781502667991 (library bound) | ISBN 9781502668004 (ebook)
Subjects: LCSH: Minerals–Juvenile literature. | Electron microscopy–Juvenile literature.
Classification: LCC QE365.2 L66 2024 | DDC 549–dc23

Editor: Jennifer Lombardo
Copyeditor: Danielle Haynes
Designer: Deanna Paternostro

The photographs in this book are used by permission and through the courtesy of: Cover bluecrayola/Shutterstock.com; p. 4 StockMediaSeller/Shutterstock.com; p. 6 Sergey Nivens/Shutterstock.com; p. 7 Papah-kah/Shutterstock.com; p. 8 Inkoly/Shutterstock.com; p. 10 Platon Anton /Shutterstock.com; p. 11 ImageBank4u/Shutterstock.com; p. 12 Jose Luis Stephens/Shutterstock.com; p. 15 Fokin Oleg/Shutterstock.com; p. 16 Kuttelvaserova Stuchelova/Shutterstock.com; p. 18 (main) Collective Arcana/Shutterstock.com; p. 18 (inset) KrimKate/Shutterstock.com; p. 19 Ryancclong photography/Shutterstock.com; p. 21 luchschenF/Shutterstock.com; p. 22 sichkarenko.com/Shutterstock.com; pp. 23, 26 Wirestock Creators/Shutterstock.com; p. 25 vsl/Shutterstock.com; p. 28 (top) Followtheflow/Shutterstock.com; p. 28 (bottom) 24K-Production/Shutterstock.com; p. 29 (top) Michael LaMonica/Shutterstock.com; p. 29 (bottom) Idutko/Shutterstock.com.

CPSIA compliance information: Batch #CSCSQ24: For further information contact Cavendish Square Publishing LLC at 1-877-980-4450.

Printed in the United States of America

CONTENTS

All the food we eat is organic material. However, when you see an "organic" label in the store, it means something different. Organic foods are grown without the use of chemicals that kill weeds and bugs.

ELEMENTS AND MINERALS

Chapter One

To scientists, there are two main types of matter: organic and inorganic. "Organic" describes anything that is or once was alive. For example, a leaf that falls off a tree is organic matter. So is the tree itself. "Inorganic" describes things that were never alive. Rocks, water, and metals are all examples of inorganic material.

Minerals are inorganic. They don't eat, **excrete**, or move. They can grow, but this growth isn't the same as organic growth. A living thing grows by taking in energy from food or sunlight. A mineral grows by **accumulating** more of the elements that make it up.

Fast Fact

Rocks are made up of various minerals. However, rocks themselves are not minerals. Think of it like a building: A building can be made of brick, glass, and steel, but it isn't any one of those things alone.

What Are Elements?

An element is a substance that can't be broken down or changed into a different substance. Each element is made of a single type of atom. Atoms are the building blocks of everything that exists. Bodies, chairs, stars—all of these and much more are made of different atoms. They can be seen only with

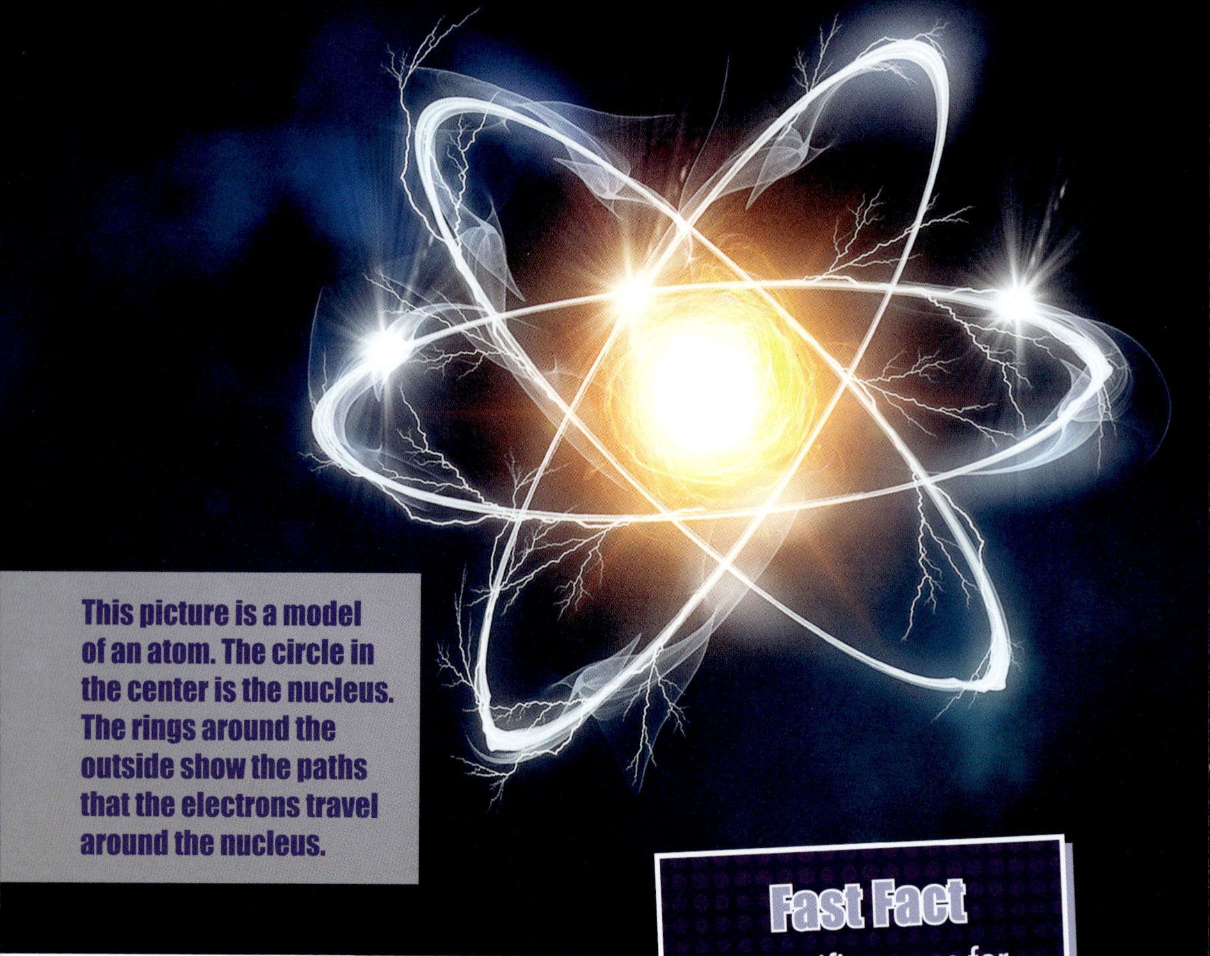

This picture is a model of an atom. The circle in the center is the nucleus. The rings around the outside show the paths that the electrons travel around the nucleus.

electron microscopes. Atoms are tiny, but they can be broken down into even smaller particles. At the center of an atom is a tiny bundle called the nucleus made up of particles called protons and neutrons. Electrons are a third type of particle. They circle the nucleus.

Fast Fact

Scientific names for minerals are based on the mineral's core elements. Sulfates are minerals containing sulfur compounds. Minerals formed from silicon are usually called silicates.

When two or more different kinds of atoms join together chemically, we call the result a compound. Minerals can be either elements or compounds. For example, gold is a mineral made entirely out of many atoms of the element of the same name. In many cases, compounds form on their own in nature. Sometimes

Quartz is an example of a mineral that's made of a compound. Each molecule of quartz contains one silicon atom and two oxygen atoms. Quartz is often used to make kitchen countertops such as the one shown here.

THE ARRANGEMENT OF ATOMS

The pattern of a mineral's atoms is the most important thing about it. This is what determines how each mineral behaves. For example, diamonds and graphite are both known as native minerals because they're made up entirely of one element. In this case, the element is carbon.

Diamond is one of the hardest substances on Earth. Graphite is one of the softest. Diamonds are clear, and graphite is black. Diamonds are highly valued and can be found in expensive jewelry. Graphite is very common; it's what forms the tips of most pencils. The only thing that makes diamonds and graphite so different is the way the carbon atoms are arranged inside them.

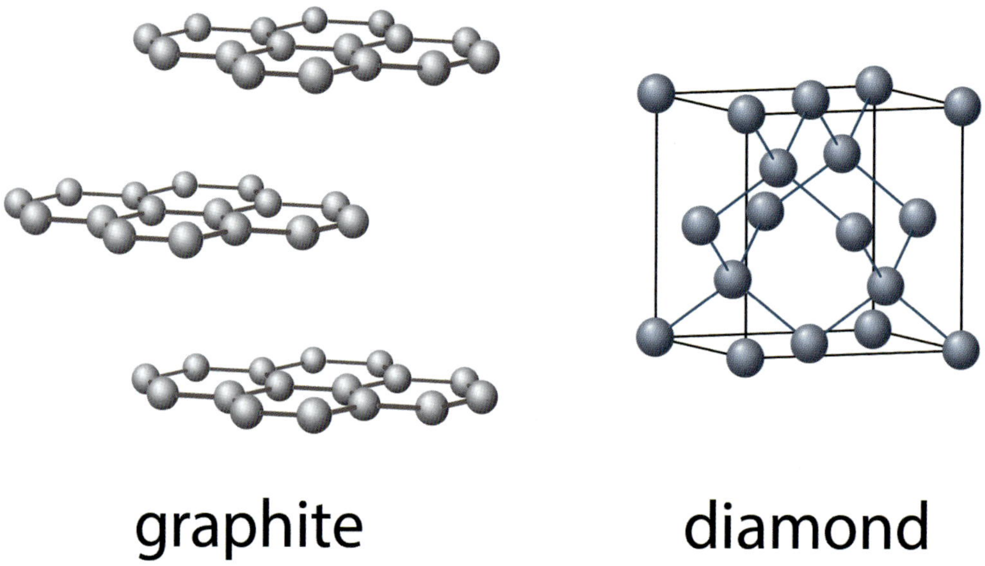

This picture shows that the atoms in graphite are arranged in layers, while the atoms in diamond are arranged in a much stronger pyramid form.

scientists can force them to form in a lab. These lab-made compounds are called "synthetic."

All the known elements are organized in a guide called the periodic table. They're arranged according to the number of protons in a single atom. The table begins with hydrogen, which has one proton per atom. Some of the elements on the periodic table are synthetic only, but most can form on their own in nature.

Fast Fact

In 1869, Russian scientist Dmitri Mendeleev arranged all the elements that were known at that time into the periodic table. He knew more would be discovered, so he left gaps to be filled in by later scientists.

What Is a Mineral?

For a substance to be considered a mineral, it must meet certain requirements.

- It must be inorganic. A mineral isn't alive.
- It must form naturally. Scientists don't consider synthetic minerals to be the same as minerals that form on their own.
- It must be a solid under natural conditions, not a liquid or gas.
- It must be crystalline. This means the atoms that make it up are arranged in a pattern. The pattern is different for each mineral.

LAYERS OF THE EARTH

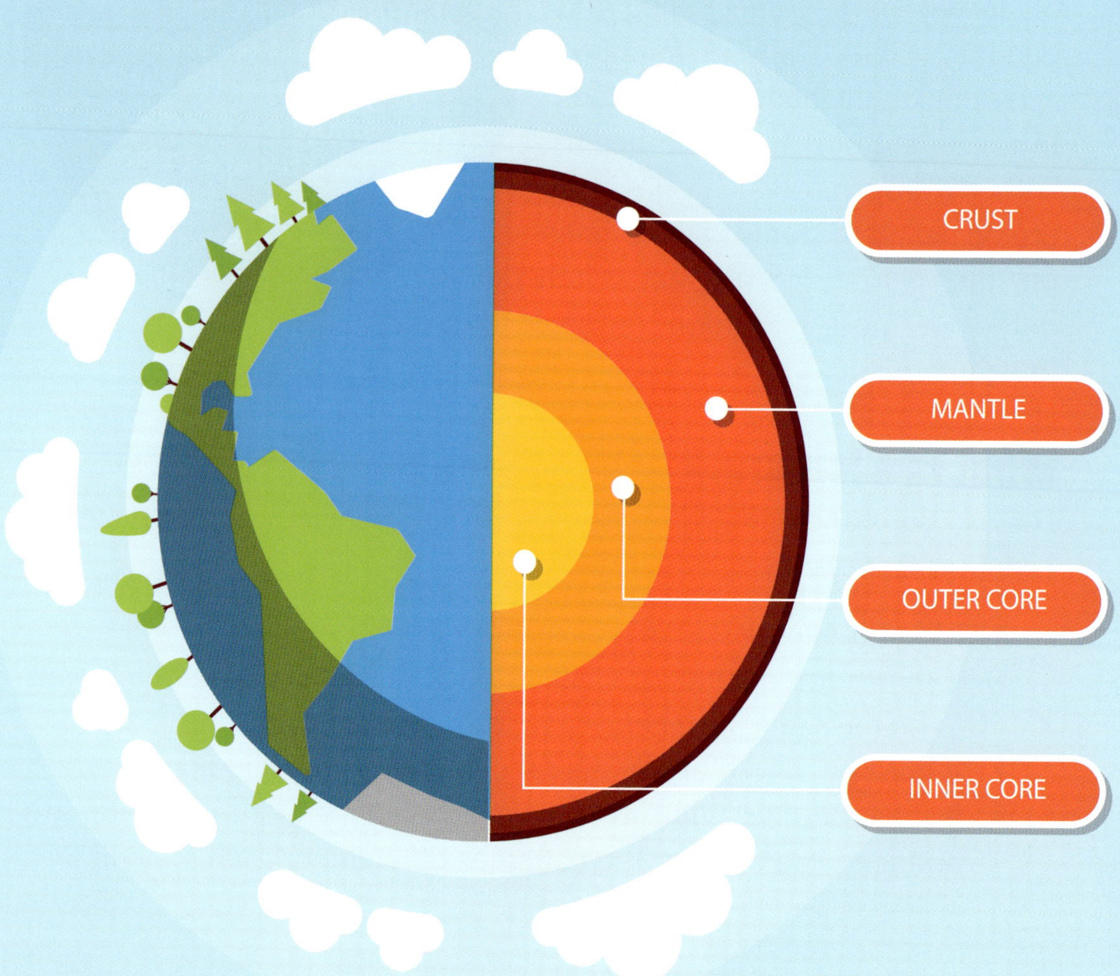

Our planet is made up of layers. The top layer, or crust, is where we find minerals. However, all four layers play a role in the formation of minerals.

Chapter Two

FINDING MINERALS

Minerals are found in Earth's crust—the top layer of solid dirt and rock that forms the surface of the planet. Earth's surface changes constantly. Sections of its crust grind and crunch against each other, exposing some layers and burying others. These forces create cracks in the rock. In many cases, liquid rock fills the cracks. It

Fast Fact

Liquid rock is called magma when it's underneath the crust and lava when it's on top of it.

Volcanic rocks have a lot of minerals in them because they are formed from lava.

isn't a mineral, but it has minerals in it. As it cools, it becomes rock and mixes with the surrounding rock.

Mining Ore

When the cracks in rock are filled with solid minerals, they're called veins. Some rocks have a lot of a certain mineral in them. The mineral can be removed from these particular rocks and used

This photo shows a gold vein in a gold and copper mine.

by people to make other things. These special rocks are called ore, and the process of removing ore is called mining. After miners remove the ore from the earth, they **refine** it to take out only the minerals.

Many native minerals exist as metals. These include copper, silver, gold, and platinum. Large formations of pure metal are rare in nature; metals usually must be refined from ore. Although they're solid at room temperature, metals can be heated to become malleable, or easily shaped and bent. This is how we can make metals into useful and decorative forms, such as jewelry or computer circuits.

Metallurgists are people who work with metals. Metallurgists often combine native metals into metal mixes called alloys to better suit some purposes. Brass and steel are good examples of metal alloys. Gold alloys are often used in jewelry because pure gold is soft and breaks easily.

Common and Uncommon

Some minerals are very easy to find. Quartz, feldspar, copper, and nickel are all very common. This makes them popular for many different things, especially building materials. Because they're so easy to find, they're inexpensive compared to some other minerals. For example, copper is used in electrical wires because it's a very good **conductor**. Gold is also a good conductor, but

Fast Fact

Some minerals in rock can also be **dissolved** in water. When the water **evaporates**, the minerals harden again. This kind of deposit is the main source of the world's salts, and it's how stalagmites and stalactites form in caves.

PROBLEMS WITH MINING

Mining is an important part of the life we live. Without it, we would have no cars, phones, bridges, or thousands of other things. However, mining can be harmful for the planet and for the people who work in mines. Mining pollutes the air, ground, and water; causes health problems; kills animals; and contributes to climate change.

Although we can't stop mining, some experts have ideas on how to slow it down. One of these ideas is to take the minerals out of old items and reuse them. For example, when a person gets a new cell phone, the minerals in their old phone can be pulled out and reused.

if we used it in wires instead of copper, items such as toasters and lamps would be much more expensive!

Other minerals are so hard to find that it has taken years to learn they even exist. Some minerals form under unusual conditions and appear as tiny crystals hidden among more common minerals.

Fast Fact

The most expensive mineral on Earth is a metal called rhodium. It's very hard to find, and as more of it keeps getting used up, it will only get more expensive. Its main use is in a car part called a catalytic converter, which reduces the air pollution cars cause.

Other minerals can't naturally form on this planet at all. In 2022, scientists used an electron microscope to look at a piece of a massive **meteorite** that landed in the country of Somalia. In the meteorite, they found two minerals never seen naturally on Earth.

Feldspar is one of the most common minerals on Earth.

When you look at the salt in a salt shaker, it might appear as a white powder. However, under a microscope, salt's crystalline structure becomes clear.

TESTING MINERALS

Chapter Three

Sometimes scientists need to test minerals. This might be because two minerals are hard to tell apart by looks alone or because a new mineral's properties are unknown. Looking at the mineral's atomic structure is one important test. As atoms bond to form minerals, the arrangement of those atoms makes a specific crystal shape. The size of the crystals depends on how quickly they formed. When lava cools quickly, the crystals that form might be visible only with a microscope. If it cools slowly, the linked crystals are large. No matter their size, the crystals' shape mirrors the arrangement of the mineral's atoms.

Fast Fact

Gems such as rubies, diamonds, and sapphires don't come out of the ground looking polished and shiny. Jewelers use special tools to shape them that way.

Breaking and Scratching

"Cleavage" and "fracture" are terms that explain how a mineral splits or breaks. Cleavage is the way a mineral naturally splits. Fracture describes how the edge of the break looks. For example, is it smooth, sharp, or crumbly?

Calcite [*top*] and quartz [*right*] can look very similar. Testing them for things such as how hard they are and how they break can help a scientist tell them apart.

Another helpful way to identify a mineral is by its hardness. This is measured by a mineral's resistance to being scratched by other substances. The Mohs Hardness Scale, which was created in the early 1800s by German scientist Friedrich Mohs, ranks minerals from 1 to 10. Higher-numbered substances

Slate's cleavage makes it naturally split into thin sheets of rock.

Fast Fact

Mineralogists might also burn a mineral's powder. The color of the flame is another clue to a mineral's identity.

scratch lower-numbered substances. Minerals with a hardness of 6 or more easily scratch glass.

The softest mineral is talc. Talc can easily be broken and scratched, so it's often used in its powdered form. The hardest mineral is diamond. Drills with diamond tips can cut through almost anything.

RADIOACTIVE MINERALS

Some minerals have a property called radioactivity. This means they give off waves of energy. We call the energy radiation. Radiation can be very harmful to a person's body, but it can be useful in other ways. For example, minerals that contain the element uranium can be used in power plants.

Radioactivity can also help scientists identify the age of the mineral they're looking at. For instance, potassium and rubidium are two weakly radioactive elements that are found in many different rocks and minerals. They always decay at the same rate, so measuring the amount of these elements in minerals can tell scientists how long ago they formed.

Color, Luster, and Streak

The same mineral often comes in several different colors, so using color alone to identify a mineral may not be accurate. For example, quartz is often clear, but **impurities** can turn it colors such as green, blue, or pink. Some minerals look dull in normal light but become colorful in **ultraviolet** (UV) light. The word for these minerals is "luminescent." Some minerals also change color when they're heated or hit.

The way in which a mineral reflects, bends, or absorbs light is known as luster. It is often more helpful in identifying a mineral than color is. Some words scientists use to describe a mineral's luster are "glassy," "oily," and "metallic."

Although minerals often can be seen with the naked eye, looking at them under a microscope can give mineralogists more information about them.

Fast Fact

Inorganic decay is different than organic decay. In radioactive minerals, decay means that they get less radioactive over time. Every radioactive element has a half-life, which is the time it takes for half of the radioactivity to decay. The half-life is different for every element.

In some cases, a mineral is a different color as a powder than it appears in rock form. Mineralogists can use a streak plate to see that color. Scraping a mineral across a streak plate leaves a line of ground mineral for observation. Streak is another important part of mineral identification.

Glass is made up mostly of silica, which can be found in sand. When lightning strikes sand, it heats up the silica and turns it into a kind of glass (*shown on page 23*).

USING MINERALS

Chapter Four

Many of the things we use and do in everyday life wouldn't exist without minerals. Windows wouldn't have glass in them without silica, soda ash, and limestone. Surgery would be impossible without sharp metal tools. All our water would have to be boiled to remove germs if we didn't have minerals such as dolomite to filter it before it comes out of the faucet.

Furthermore, we need certain minerals in our body to keep us healthy. For example, people who don't have enough of the mineral iron in their blood have a health condition called anemia. This can make them dizzy and tired.

Fast Fact

Some kinds of minerals have unusual properties. For example, magnetite is a mineral formed from the elements iron and oxygen. People used it in early compasses because it's a natural magnet.

Vitamins and Minerals

Inorganic matter is very different from organic matter. However, plants and animals can't live without minerals. Plants need certain substances called nutrients to grow and stay healthy. Many of these nutrients are minerals that are found in or added to the soil where plants grow. The minerals dissolve in groundwater and are absorbed by plants' roots. When people and animals eat plants, their bodies take in the minerals inside the plants. People can also get these minerals by eating animals that eat plants.

Fast Fact

Petrified wood is organic tree matter that has turned to stone! Water with dissolved minerals in it breaks down the wood. When the water evaporates, the minerals are left behind.

Many people confuse minerals with vitamins, but they aren't the same. Our bodies need all vitamins, but only some minerals. The key difference between vitamins and minerals is that vitamins are organic substances and minerals are inorganic. Our body can make some of the vitamins we need, but it can't make any of the minerals; we have to get them all from our food.

Doctors recommend getting a certain amount of each mineral, and this amount depends on how much of it the body needs. Getting too much or too little of a mineral can hurt the body and keep it from working properly. Too much salt, for instance, can cause high blood pressure. Getting too much of other minerals can make a person feel sick. Some other minerals, such as lead, can be dangerous or even deadly if eaten.

Iodine is one of the trace minerals, which means we need only a tiny bit of it. It helps the body make chemicals called hormones that keep it running properly. Seafood is a good source of iodine. Also, some kinds of salt have iodine added to help people get enough of it in their diet. This is called iodized salt.

MINERALS IN BONES

The mineral calcium plays a major role in the human body. It helps form teeth and bones. In fact, calcium can make up as much as 2 percent of a person's body weight! About 40 percent of bone weight is calcium. The rest is an organic material called collagen.

After death, a body decays, or breaks down. Microscopic creatures such as bacteria play a large role in this process. However, these creatures can't eat minerals. The calcium in bones makes it hard for them to get at the collagen. This is one reason why skeletons resist decay much longer than the rest of the body.

Bones decay much slower in dry places, such as deserts, than they do in wet ones, such as forests. This is because the creatures that break down organic material need water to live.

Mandatory Minerals

Minerals are an important part of all life on Earth. They have given us many of the inventions that make our lives comfortable and are also essential for keeping us alive and healthy. Studying minerals, including under a microscope, can teach scientists a lot about our planet. For example, knowing which conditions certain minerals form under tells scientists what the planet was like thousands or millions of years ago in spots where those minerals are found. It can also tell them more about the solar system. For instance, the mineral hematite has been found on Mars. Since it only forms near water, this tells us that Mars once had liquid water on its surface.

Fast Fact

The mineral olivine can form only under the surface of the earth, deeper than 43 miles (70 kilometers). Finding olivine in the rocks around Hawai'ian volcanoes tells us the volcanoes are very deep.

Getting a close look at minerals allows scientists to learn many things about the world around us—and the world beyond our planet. Even the tiniest details found under a microscope can lead to a big discovery!

THINK ABOUT IT!

1. Look at the things in the room around you. Which ones are organic and which are inorganic?

2. Why do you think some minerals don't exist on Earth?

3. Why is it important for scientists to know which mineral is which?

4. What other everyday uses for minerals can you think of?

GLOSSARY

accumulate: To gather or pile up, especially little by little.

conductor: A material that can allow electricity, heat, or sound to pass through it.

dissolve: To break down a solid when a liquid mixes with that solid.

electron: An elementary particle that has a negative charge of electricity and travels around the nucleus of an atom.

evaporate: To change from a liquid to a gas.

excrete: To separate and remove waste from a living body.

impurity: Something unwanted that is mixed in with a substance.

meteorite: A space rock that has reached Earth's surface.

molecule: The smallest part of a substance that still has all the characteristics of that substance.

refine: To produce a purer form of something by removing unwanted substances.

ultraviolet: A range of wavelengths in light beyond the violet end of the visible color sequence.

FIND OUT MORE

Books

Martin, Claudia. *Minerals*. Lake Forest, CA: Quarto Library, 2018.

Owen, Ruth. *Minerals*. Minneapolis, MN: Bearport Publishing, 2022.

Pettiford, Rebecca. *Minerals*. Minneapolis, MN: Pogo, 2019.

Websites

KidsHealth: Minerals
kidshealth.org/en/kids/minerals.html
Learn more about some of the minerals that help keep us healthy.

Mineralogy4Kids
min4kids.org
Play games, identify minerals, and read about the rock cycle that helps create them.

University of Waterloo Earth Sciences Museum: Minerals for Kids
uwaterloo.ca/earth-sciences-museum/resources/just-kids/minerals-kids
Read about specific minerals and their properties.

Publisher's note to educators and parents: Our editors have carefully reviewed these websites to ensure that they are suitable for students. Many websites change frequently, however, and we cannot guarantee that a site's future contents will continue to meet our high standards of quality and educational value. Be advised that students should be closely supervised whenever they access the internet.

INDEX